ELECTROMASS
THE SAME PRINCIPLES AT EVER
BY JUSTIN SANDBURG

Cover photos taken from The Space Shuttle Endeavour,
August 2007

Revised: 4 July 2010

CONTENTS

ELECTROMASS

The Same Principles at Every Scale

Justin Sandburg

iUniverse, Inc.
New York   Bloomington

ELECTROMASS
The Same Principles at Every Scale

iUniverse books may be ordered through booksellers or by contacting:

iUniverse
1663 Liberty Drive
Bloomington, IN 47403
www.iuniverse.com
1-800-Authors (1-800-288-4677)

Because of the dynamic nature of the Internet, any Web addresses or links contained in this book may have changed since publication and may no longer be valid.

ISBN: 978-1-4401-9379-8 (sc)
ISBN: 978-1-4401-9380-4 (ebk)

Printed in the United States of America

iUniverse rev. date: 09/24/2010

Cover photos taken by the Space Shuttle Endeavor, August 2007

A Curious Observer:

"Even the most distant pictures I see, it looks like electrical behavior repeating itself. What is the astronomy take on electricity?"

The *standard astronomical model* follows the expanding universe decision paramount with Hubble (1929) and Tolman (1930). That same model follows the mathematics of Einstein (1905) and combines it with the *required* supermassiveness of black holes to date The Universe. The model also follows the mathematics of Shrodinger et al (Quantum Mechanics), and Feynman et al (Quantum Electrodynamics), which explains the atom's behavior from an observational standpoint only. This becomes immediately cumbersome when complex protein chains start FOLDING several different ways, beginning your personalized archetype. Everyone is looking for a model that explains the why, because fundamentally these different sets of mathematics do not agree with each other; none are reasonably explained; and all require the introduction of new and unconfirmed fundamentals. *"Despite the general importance of plasma in astrophysics, the standard model continues to delineate that electromagnetic forces are not important at large cosmological distances. The reason for this is generally believed that unlike the other three forces which are attractive only, electromagnetism is both attractive and repulsive and over large cosmological distances, electromagnetic forces are believed to cancel each other."* -Wikipedia

Most astronomers are interested in the majestic shapes and awesomeness of the array. They labor continually on the arduous task to make precise observations, which coincide with provided mathematics. However, math doesn't prove concept. Math translates data. The scientific method demands that an order be placed so conclusions cannot be validated by hypotheses, and so mathematical tools cannot validate impossible ideas. First the concept; then the mathematics become evident; then and only once all attempts to disprove it are exhausted, does a concept become theory. It still requires years before a theory deserves a capital letter, and even longer before it becomes Law. Each field of science makes assumptions—putting limits of scope or validity. Such logic and limited data received demands the option for something new. Since we have *certain* data received, conclusions can now be drawn for the first time.

The scientific method is slightly different than what is observed in things like traffic law or the endless array of taxes, fees, rules and regulations. I'm more pro-government than most, but I tend to question the Laws of Physics (like a limit on the speed of light), alongside other laws like speed enforcement vs. lane enforcement. I don't condone speeding in school zones but I also don't condone slow driving in the left lanes because it's as unnatural as a slower flow in the center of a river.

Although it's heart demands answers to many philosophical implications, this book focuses on the strongpoints of Physics—attempting to refrain from expressing feelings one may have when hit with this concept—"feelings," says my long favorite Pink Floyd's <u>The Trial</u>, "of an almost human nature. This will not do!"

PRINCIPLES:

After withstanding 1,400 years, the out-dated theory that the Earth stood at the center of the Universe and other objects go around it was replaced by the heliocentric model of the solar system—first rationalized by Nicolaus Copernicus in 1497, then published in 1543, initiating the Scientific Revolution. The geocentric model was usually combined with a spherical Earth by ancient Greek and medieval philosophers. The ancient Greeks also believed that the motions of the planets were circular and not elliptical; a view that was not challenged in western culture before the 17th century. The geocentric model held its support into the early modern age; from the late 16th century onward it was gradually replaced by the heliocentric model of Copernicus and Galileo. Consider the odds; as of 1480, there were still rumors that the Earth was flat.

Isaac Newton (1642-1727) was first published in 1687, with the *Philosophiæ Naturalis Principia Mathematica,* where he spells out the Laws of Motion as well as Universal Gravitation. Newton is still acclaimed as the most influential human scientist to date. Newtonian dynamics was modified for orbital motion by Johannes Kepler, and the geocentric model was put to bed.

In the 1780's, Charles Augustin de Coulomb was able to create the laws to describe charge separation and electrostatic fields. In 1835, Carl Friedrich Gauss developed Laws of Magnetism, closely tied to Coulomb's Law, but was not published until 1867. James Clerk Maxwell made an important correlation with André-Marie Ampere's Circuital Law in his 1861 paper, showing that a magnetic field generated by a flow of current, is exactly identical to a flow of current generated by a magnetic field. Maxwell is given the credit for the combined ideas of Faraday[A], Gauss, Coulomb, Lorentz, Ampere, Poisson, although Oliver Heaviside and Willard Gibbs combined the laws into a distinct set of four equations in 1884, called Maxwell's equations.

On the astronomical front in the early 1900's, telescoping technology helped astronomers view much more distant objects. It was then that they observed a phenomenon called "red shift," (discussed in the chapter on cosmology) and they were quick to nail down its cause. Unfortunately over the past one hundred years, they have been fighting diligently to keep their early twentieth century predictions valid. Hilton Ratcliffe[8] has organized the history very well in his confessions of a dissident astronomer, "The Virtue of Heresy." He goes into much more detail than I have, and his book is a highly recommended read to understand what actually did happen in the last hundred years, and why. Now I'm not much for conspiracy theories, nor do I venture into the notions of scientific corruption in the peer review system. I tend to believe that perhaps some parts of Physics have been classified for good reason. Some of the greatest parts of this enrichment program are not in rocket science or missile technology, but rather in health care, energy, botany, climate control, visualizations, and space exploration.

The accepted models have been interesting to approach, and have been extremely successful. This book shows dissent from widely accepted astronomy and offers a change of mindset that doesn't rely on either of two sets of Physics; twisting basic principles which justify and expand each other. I believe science to describe everything—living or dead—from human reproduction to the life-cycle of a star. Any model that is contradicted by another model is not an accurate tool for Nature; only for a certain scope.

During high school Physics class, you are taught the basic principles of projectile motion and electricity. This is a solid base for physics students; however, the math that is used to describe the phenomenon is only accurate if certain things are ignored. In the first year of college, students are taught calculus based physics. This makes sense of the equations because the math used in calculus is what created the formulas. Electricity is also explained using calculus. In the second year, the math is replaced once again by differential equations, to stop ignoring things like wind resistance (drag). In the third year, electricity and magnetism is explained using a slew of different mathematics, partial differential equations, matrices and linear algebra, calculus, as well as some additional operators to multiplication, addition, division, and subtraction—dot product and cross product. Physics is taught that way because too many students would be caught up in the math required for the third year; they would miss the point of Physics along the way. So they teach it over and over again, each time using harder and harder math or including things once ignored. The *history* of Physics is no different than how it is taught.

Physicists fight to know what to ignore in every situation, because we are held by the math we know. At every advancement, the mathematics learned were consequences of forward thinking physicists who inspired the required mathematics to be realized. One mistake in the last hundred years is that the physics left "ignored," stayed ignored and accepted mathematics took over. This gave rise to a thought process of exactness and correctness, deciding things are proven. That's how math works. But Physics allows for two different phenomena to be the cause of the same observation. It is taught to include proper variables despite the math, not as a consequence of the math.

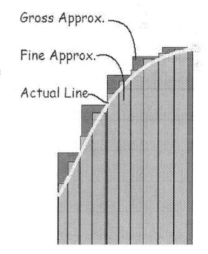

It will be necessary to discuss a wide range in scales. We will visit the very small, the hand held, the visible, and the largest structures known. We'll also discuss the very distant and the very bizarre. To keep it tidy, it's best to break the scales down by name: Atomic, Standard, Planetary, and Galactic. The same rules apply, but in different amounts of mass and charge. This is the basis of *Electromass*.

THE BASICS:

Separated mass attracts—called Newton's Law of Universal Gravitation. Simple rules of separated charge: Opposites attract, likes repel; called Coulomb's Law. Both Universal Gravitation and Coulomb's Law hold.

Light is represented as waves. The wavelengths can stretch from very fast moving, very short wavelengths, to waves that are kilometers long before they repeat. It is a set of possible wavelengths through which energy can be transmitted. What our eyes see as visible is a sliver of possible wavelengths of The Electromagnetic Spectrum.

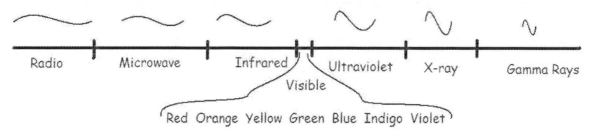

Besides the spectrum, there are limited other things that exist. That's how powerful the spectrum is. Those other things in their most basic form are *Protons, Neutrons, and Electrons.* It is necessary to be rigorous and detail the known. These little points are things are of little concern to most. Even the experts get lost with the entire math buried behind these tiny properties of mass and charge. They are, however, important and the scope should never be overlooked.

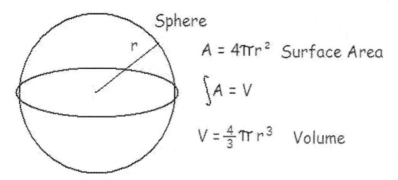

The mass of a *proton* is $1.67262158 \times 10^{-27}$ kilograms. Its charge is $1.602\ 176\ 487(40) \times 10^{-19}$ Coulombs. The diameter of a proton is 10^{-15} m. Its radius is 5×10^{-16} m. If a proton is a sphere, calculate its volume.

A proton all by itself does not emit. It does not energize. It can be influenced by its surroundings, but otherwise, it remains just a positive charge on a massive body; call it "a capacitor" for its ability to hold a charge.

The *electron* was first theorized by Richard Laming (1838–51), and discovered by J.J. Thomson (1897). Many steps have been made in understanding it. The charge of an electron is negative (-)1.602 176 487(40) $\times 10^{-19}$ Coulombs. The mass of an electron is $^1/_{1836}$ that of a proton, weighing in at 9.10938188 $\times 10^{-31}$ kilograms. *The enclosed (numbers) are the uncertainty in the last digits of the decimal.*

The diameter of an electron is unknown. In fact, it is widely accepted that electrons cannot be even considered as little ball bearings, but rather only as energy shells, dictated by Electricity and Magnetism and an "atomic weak force" in the nucleus. A great struggle has been made to determine if the electron is a particle or a wave, because two experiments can show conclusively that it is both.

It seems reasonable that since electrons have so little mass, they are very much dictated by their charge, constantly seeking protons that will equalize, or rather, decrease their instability. They are energy waiting to be harnessed. Alone, electrons will shake violently <u>and</u> travel along their projected path. If a proton is near its projected path, the electron will bend towards that proton, just like you would expect from a particle. The proton, like a football center, will hold its ground due to its "inertia," (Newton's First Law). Conversely, if an electron is shot at two parallel strips, its energy will be observed through both strips. This is the wave-particle duality for electrons. For more on these two properties, search Crooks tube and Young's double slit experiments.

We know that protons are a charged mass in the nucleus (center) of an atom. We know that electrons are the dichotomy—a negative charge with a 40% uncertainty on the ninth digit that says pretty much equal, and opposite—charge. They attract. They collide. If each is traveling the right speed with respect to the other, the electron can be sent into orbit, "bounding" the proton—so we have Hydrogen (neutral). Several tests of shooting electrons and protons, and combinations of electrons and protons together are in the works. Other tests of shooting just the electrons in orbit have recreated a situation with the proton by itself once again. Very interesting work in the field of High Energy Particle Physics has been limited to removing both electrons from a Helium atom. The Large Hadron Collider is funded by and built in collaboration with over 10,000 scientists and engineers from over 100 countries as well as hundreds of universities and laboratories. Producing atoms may require protons and electrons to both be moving at precise vectors, not by directly colliding one into the other.

The Hydrogen (neutral) atom is a Proton with an Electron. Its emission (light) spectrum looks like the graph to the right.

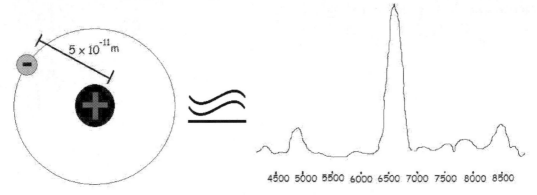

That graph may look like a mess, but it's just a line. It's a signature for that element's structure. Sometimes signatures look messy, but they have an underlying meaning behind them. My fingerprints imply me; I imply my fingerprints. They are interchangeable. To some equation, the two pictures above can be made equal.

The electron is so fast, the entire scientific community (in chemistry as well as physics) has been forced to use Quantum Mechanics (QM)—some math to predict angular momentum or position of the electron. From the descriptions in QM, scientists have found that electrons make up exciting patterns called energy shells or energy levels. We can to some degree of Heisenberg's uncertainty, draw atoms. For Hydrogen (neutral) there is a high probability that the electron will be 5×10^{-11}m away from the nucleus. The probability varies very little from this radius. Hydrogen (neutral) is somewhat stable, except angular momentum is not conserved. The electron doesn't have another electron to spin down when it spins up, so you get an element that is highly likely to bang into and react with another atom like an Olympic Hammer Throw.

HAND RULES:

Take a moment to look at something a little more tangible; however, it is valid at any scale. Iron conducts electricity well, and it is cosmically the most abundant element found. Let's talk about Iron rods (or wire). If Iron could be lined up one atom in diameter, this could also be called an Iron rod. Any size is fine. When electricity flows down that rod, it creates a magnetic field. Point your right hand thumb in the direction of the flow, and your fingers curl in the direction of the magnetic loop.

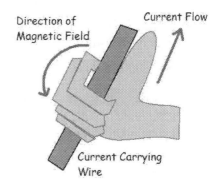

If you sent a current of one electron per second down the wire, the magnetic field would look like a corkscrew, with the field regenerating every second. If you sent a steady current, the magnetic field would look like a cylinder with the rod 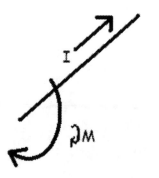 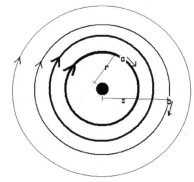 at its center. It's important to clear something up; the fields' equal strength areas have a shape but not a structure. It is actually a series of edgeless cylinders—one on top of the next—each bigger and losing strength. Imagine the general shape of the field, but pick points to observe. Point "a" a distance "r" from the rod, has a field is perpendicular to the direction of the flow of current; and a direction associated with it (look at your right hand). Now pick a point "b" a distance "s" away, and observe the field there. No matter how quickly the magnetic field reduces, it can never be ignored.

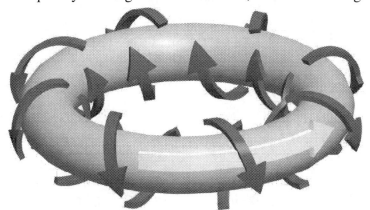

Say that rod was curved into a circle, and say current could keep going around it; the magnetic field would look like a torus. Just like the rod, the strength of the magnetic field decreases as you get further from the source (don't forget the associated direction); but it's not just a torus.

Remember the "cylinder?" It's a continuous field. So what happens at the center of this ring of flowing charge? Well, it turns out that when they meet, magnetic field lines "align" themselves based on field strength, by vector addition. So the torus flattens out and you have nearly parallel lines at the center of the ring. Part of Gauss' Magnetic Field Law is the divergence theorem. It specifies that the magnetic field in and out of any enclosed reference is zero. So those nearly parallel lines that run straight up eventually curve, and eventually come back to the starting point; a bigger, weaker torus. The standard model pusher invented magnetic disconnection and reconnection, which violates this Theorem. Sunspots don't always have continuous rings of plasma like what you see on the first page. But the ejection is steered and corralled by these never breaking magnetic fields.

Now if instead of a flow around that ring, think of the ring itself being charged, and the ring is spinning. You still have a flow of charge with respect to an observer, and the magnetic field would look identical. If you were to take a set of rings that fit inside each other, and there were enough to stack on top of one another, you could make a sphere of charged rings, all with the same axis of rotation.

Zoom out, and those field lines eventually have room to make yet another "torus." Superglue the knot of an inflated balloon to the top of the inflated balloon; you would have a pretty good representation of the knot's magnetic field at great distances.

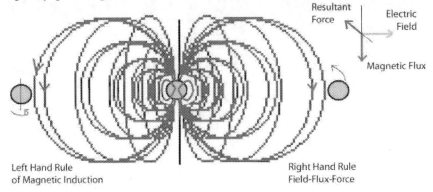

The magnetic field in its direction can induce a spin <u>and</u> a tangential force on a nearby object. If a spinning charge is near a similar charge, it will cause the other charge to spin by way of the "left hand rule of magnetic induction." Point your left thumb in the direction of the magnetic field down the axis of the second object, and the fingers show the induced spin on it. The other rule is to draw the directions of the Field-Flux-Force: The Electric Field, the Magnetic Flux, and the resultant Force, by holding your right thumb and first two fingers, mutually perpendicular to each other (compliments of Michael Faraday[A] (1840), Hanns Alfven, 1937).

A "NEW" LOOK:

Now that we have explained what matters, drift your attention to what is observed at the atomic scale, and how to explain it. Niels Bohr (7 October 1885 – 18 November 1962) was a Danish physicist who made fundamental contributions to understanding atomic structure. Bohr described the Hydrogen atom as what is today referred to as a Bohr magnetron, with a negative electron traveling in a near circular orbit around a central positive nucleus as a "*single turn endless electrical coil,*" which produces an almost constantly directed magnetic field. Bohr's model eventually led to failure and any

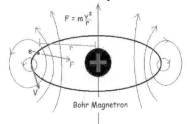

profound scientific thinking was muted by the predictive power of Quantum Mechanics. But in my mind, the metaphysical questions Bohr was attempting to answer needed to be considered in any physical thought and QM does not do that, no matter how exquisitely accurate it is at atomic scales. Math is fun, but it has to make physical sense.

Here we have Bohr's model of the most fundamental element of the periodic table—the Hydrogen Atom. The Bohr magnetron was dismissed because none of the other elements in the periodic table made sense. QM had shown these weird energy shells and it did not go along with Bohr's ideas, so they were rejected. But if we take a second look, there's something that Bohr missed, which is why his model was never realized.

electromass –10

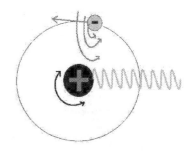

Here, the Hydrogen Atom showing an alternative site for the emission (Traditionally, it is believed that the emission comes directly from the electron changing orbital radius). The magnetic field from the electron will cause the proton to spin via magnetic induction. Since the proton has mass, it is constantly "forced" to roll. This is an important combination of mass and charge enclosed. The concept of force is trademark of Newton, whose first law says that an object at rest will tend to stay at rest. Since the Proton has mass, it "tends" to stay at rest.

It "tends" to the center of the atom, because the center is where the electron has the least influence on it. Keep in mind that since the proton has so much more mass than the electron, the electron would do most of the work "tending" *itself* to fly in this orbit. In their path, the electrons move up (to a higher orbit) which requires input of energy; they move down and there is a release of a photon, called "light" in the form of emission lines. What remains of the Proton's rolling and the electrons changing orbital radius is given off as an emission spectrum pictured and diagramed to the left. Each atomic element has its own signature of lines. They occur as a bar graph, and each bar represents a different color, or wavelength. Most humans see reflections of light off of an object, and those reflections occur at specific wavelengths (colors). If you dare look directly into the Sun, you see a blend of light blazing down on you—of many portions of the spectrum, including visible.

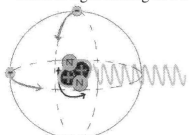

A Neutron has similar mass as the Proton, but holds no charge; call it a "dielectric." Neutrons are the final fundamental particle. They are found in every element larger than Hydrogen—bound to the protons in the nucleus—a requirement for those elements' survival.

The next atomic element is Helium, which consists of 2 protons, 2 electrons, and 2 neutrons. The two electrons trap the protons; the neutrons follow the protons' mass, and both are forced to the center. Contrary to popular belief in a "strong nuclear force," suppose as soon as nucleons touch each other, the charge shifts to the surface around the touching protons. As a dielectric-capacitor grouping, the positive charge flows over the surface of the protons, through the dielectric. While the charge is passing through the neutrons, the protons are temporarily relieved of duty. This relief oscillation creates the observation of a nucleus being held together by a strong force. When a proton is caused to escape the nucleus, it requires much more energy because it would require the positive charge to choose and fill the escaping proton. When a Neutron is caused to escape, it forces the nucleus to hold more charge for the surface. The nucleus would be maintained as a shell of positive charge on the surface, with capacitating and dielectric nucleons inside. This makes sense electrically because charged capacitors will repel until they touch, but then it takes even more work to remove them. If we can explain the strong force it may not need a name.

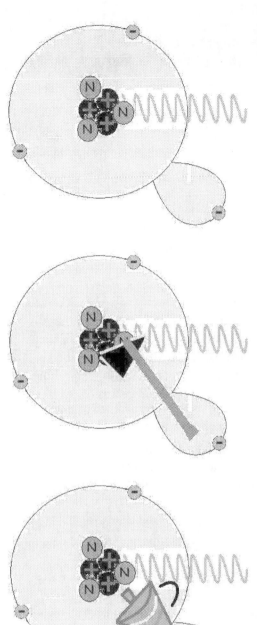

The Electrons on the Helium atom complete the first energy shell. After Helium, there is no more room for electrons on that first spherical shell. Those extra valence electrons are attracted to the nucleus but are forced to take on wider paths. Lithium has one more proton, neutron, and electron than Helium. That extra electron is in a constant battle to gain admittance into the first shell, and then is forced away by those two stable electrons. This creates an arm-like energy level and the total magnetic field makes another unique spectral signature from the rolling of the nucleus. This wreaks havoc for the conservation of angular momentum. The extra charged arm on the otherwise stable nucleus makes the atom resemble a top heavy axe being thrown at a renaissance fair, and whichever electron gets forced to the next shell would get whipped around the nucleus; unless, the whole "axe" was spinning "axially."

Being whipped around would induce an unnecessary magnetic field "rolling and shaking" on the nucleus; it would react with an overall rotation. The whole thing spins axially to that extra electron whenever possible, but that electron is so frisky, the other two electrons act to balance it and keep it away from the first shell.

Lithium (left), plus another neutron, proton and electron, gains stability with another axe handle on the north end of rotation to make Beryllium (4 protons, 4 neutrons, 4 electrons). The next many pairs gain stability just the same way, with the nucleus rotating.

From a distance there is a net zero charge, but the difference between is the reason the electron orbits. The weak force is a balance between total magnetic field holding and the capacitance of the nucleus. This is a fundamental relationship between mass rolling, capacitance charge, total magnetic field causing the rolling, and resultant emission. The weak force is not unexplainable.

Below: Ionization energies (the energy required to cause an electron to change orbit) shows how filled energy shells gain stability (reduce angular momentum). Naturally occurring abundances of Iron (Fe), Silicon (Si), Nickel (Ni), Sulfur (S), Oxygen (O), show that these paired elements have greater abundance in the solar system. As energy shells become filled, the momentum of the atom becomes balanced like a motorcycle tire.[19].

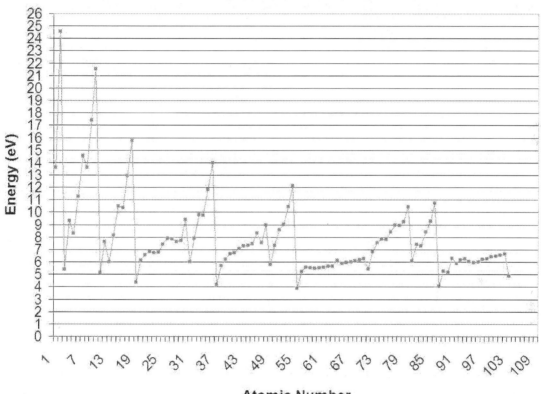

Ionization Energy for a Single Electron per Atomic Element

BEYOND THE TABLE:

Quantum Mechanics was successful in drawing every atom in the Periodic Table of the Elements. As we grow in atomic element, the electron shells get further from the nucleus, and therefore, have less of an influence on it. Conversely, many electrons have more total magnetic field acting on the nucleus. It's easier to strip fundamental protons and neutrons from larger elements. This is a phenomenon referred to as radioactive decay, and is traditionally explained by the *weak nuclear force*. Minimizing forces is called empirical science, or Occam's Razor.

electromass –13

Radioactive decay is so accurate it can determine the length of time it will take to reduce every fundamental nuclide (Isotope and ion for each element). For example, Nitrogen is found 99.635% with seven neutrons. But it is also found 0.365% with 8 neutrons. The time it takes for Nitrogen to lose that extra neutron is precisely determined. Radioactive dating is used on virtually every scale we know of to determine timeframes. Geologists have been able to draw a map that traces the evolution of earth. Oliver Manuel has taken radioactive dating to a new level with his work on the composition of the sun, moon, planets, and meteorites, all of which point to a birth ~5 billion years ago[B] (discussed in the chapter on the proposed life-cycle of a star).

Adenine   Thymine   Cytosine   Guanine

The story with Quantum goes on. Richard P. Feynman introduced Quantum Electro-Dynamics, QED. In it, he begins to correlate light with Quantum Mechanics. This is tedious work done by making many observations of light reflecting off of glass, and predicting probabilities of such reflection. The hydrogen atom alone takes several pages of handwritten labor to model and describe, but QED uses some math tricks to make certain observations quicker. It still doesn't explain them. To really understand the scope of this, one would have to take on a degree in Physics and then Biochemistry at a University. Summarizing a few key points; it should be noted that I found these numbers on the internet. The human DNA is made up of four types of nucleotide pairs: Adenine—thymine, cytosine—guanine. They tie to one another like rungs of a ladder between two helical strands of a line of sugar-phosphates. There are about 3 billion nucleotide pairs packed away in a human DNA double helix. Guanine itself has a molecular formula $C_5H_5N_5O$. Remember how just developing an equation for the Hydrogen element took several pages of mathematics?

Each human cell contains a nucleus with forty-six chromosomes. Each of these chromosomes is comprised of between 30,000 and 50,000 genes and intervening sequences. Each gene is further made up of variably sized DNA sequences. Our bodies manufacture over 100,000 proteins to differentiate all the parts of a body. It's estimated that there are about 12 billion molecules assembled for the DNA in each cell[3]. Worse yet, it is estimated that the average adult human body contains about 10 trillion cells[4]. There are 7.0×10^{15} atoms per human cell[2]. So a human body would have 7.0×10^{27} atoms[1], each taking several pages of mathematics. Supercomputers are just now able to develop models of the simplest proteins. The chances of waiting for computers to allow Quantum Mechanics to fulfill the questions of Nature are predicatively low. The closest we are to understanding the human body is by using Quantum Bio-Feedback. This cutting edge science measures and graphs resonance in the body to determine which processes are lacking. This forty-dollar test is widely misunderstood as witchcraft or pseudo-science. But one by one, doctors and patients alike are taking heed to their own "emission." The implications of this have a direct impact on healthcare for humans. If we can draw a better equation, this test immediately goes from "the graph looks similar to the flu virus," to a precise method for determining illness, and developing holistic approaches to rebuilding organ function to enhance health.

ELECTROMASS:

As observers, humans have been studying orbital motion ever since we learned that things took on orbits. We can track the orbits of planets, and even stars in our own galaxy. We witness the perturbations of them interacting. Experts have decided that since electricity has no place in the cosmos, then all the forces are due to attraction of mass "alone." This understanding plays a critical role in answering *why* astronomers have resorted to things like "supermassive" black holes. After that, they had to invent other forces like dark energy and anti-gravity. This is a serious problem to have invented some forces and neglected others at will.

The forces due to like or unlike charges may at times be "dwarfed" by attraction of mass; but it does not give anybody the right to discount that force. A simple recalculation could adjust for the calculated mass-charge of a heavenly body, and the same observations would be observed.

At any scale, based on our rules for gravity and the Coulomb Field, we can determine the radius where two charged-masses will stabilize. They are both fields that vary by the distance between the objects, squared. The two objects below orbit each other. Arbitrarily call them both positive charges.

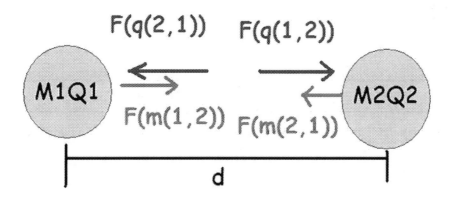

There is a force due to the charge of Q1, acting on Q2; and also a force due to the charge of Q2, and acting on Q1. Then there is an attraction of the two masses. Suppose the forces due to charge are much higher than the attraction of mass. These two objects would separate quickly and they may not orbit each other at all (just like two protons which never orbit each other without binding electrons). But if the repulsion due to like charges was smaller than the attraction of mass, they would continue to orbit each other. Their spin also plays an important role creating magnetic fields that bind them together. This will be discussed over the next several pages at a larger scale.

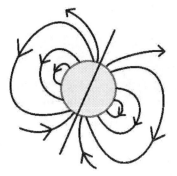

The Sun is a vivid example of a charged mass. It shows a nice line between planetary scales held mostly to attraction of mass and galactic scales, held together by something more than just mass. Our Sun is a very good source of information in understanding "population I stars."

Three chemical reactions with regard to the energies in the sun have been compiled. These energies (in Mega-electron-volts) are responsible for the sun's heat and luminosity[9]. That energy travels along the wavelengths of the Electromagnetic Spectrum, which burn our skin and brighten our skies [Table 16, 16C]. But also, large quantities of positive ions are observed indicating the sun is highly positively charged ($3x10^{43}$ H+ escaping in the solar wind per year, Oliver Manuel[9].) This makes sense because the ionization energy of a Hydrogen atom is 13.7 electron volts. So it is reasonable that the sun being so hot means that electrons do not bound the protons. Ralph Jeurgens was onto this in the 1970's. Oliver Manuel showed that Neutron repulsion at the center was responsible for a large portion of the heat. The electric sun model is close, but it concludes that all of the heat is electrical, discarding nuclear reactions and neutron repulsion—and suggests that the sun's energy is exterior, entirely from electric arcs. Electromagnetic Theory explains why proton escape (and acceleration) exists. But combining the electric sun and the neutron repulsion sun can reasonably explain everything we see, including the many atomic signatures. Groups of neutrons and protons are forced to roll, and give off signatures imitating known elements. This does not mean that those elements are separated and inclusive; just that certain groups of protons and neutrons roll well together, and as you get closer to the surface, act like boaters being thrown from a turning boat. From this concept, we should consider reversing the hydrogen-into-helium fusion model, thereby discarding the predicted life cycle of our own sun, and embrace electromagnetic theory (differential physics) on a large scale--while including mass. It actually helps to explain the rest of the phenomenon; but the math is yet to exist.

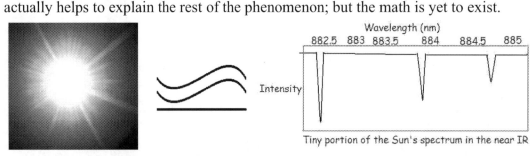

Tiny portion of the Sun's spectrum in the near IR

Just like how a Hydrogen atom has an equivalence equation to its emission, the geometry of the solar contents is somehow equal to the spectrum received. There exists an equation that will make the spectrum and the picture equal, and it must include charge and mass. The freehand drawing of the spectrum shows how spectrometers currently identify composition by recognition or similarity to patterns made by known elements.

Here we have a typical case where two charged objects are spinning in the same direction. Let's call them both positive charges. Both charges have mass, and will attract by Gravity. But also, both charges have similar charge, and will repel by Coulomb's Law. Those forces will counteract each other and a balance distance between them will be reached. Because of their spin, the magnetic field lines for both objects leave the north poles and arrive at the south poles. In an environment full of charge separation, negative charges (Electrons) from a distance will see this net positive charge and migrate towards it. When they get close enough to feel the magnetic fields, they react tangentially to the magnetic field and as seen below, brought through at the center and away. Stars align themselves with the magnetic field they are sharing, being pulled around not only by gravitational attraction, but also by following a shared magnetic field causing the main net flow through the center. Each star's spin creates its own magnetic field that would help continue the same flow—through the center around and back. Because the magnetic field is directed downward in-between the two point charges, this is the path according to the right hand rule, "Field-Flux-Force." Since we are aware of electromagnetic theory and its rules, we can draw this image and we can see that there is no supermassive center; but rather, a flow of charge through the center. This flow creates a binding magnetic field.

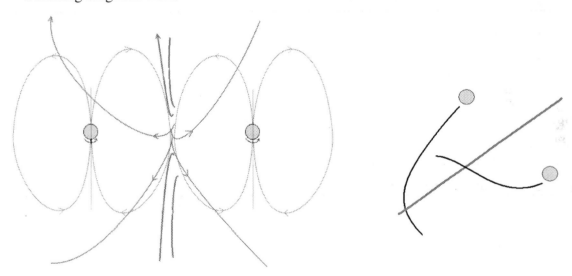

Just like a current flowing through a wire, a secondary magnetic field binds the two point charges. The circular field puts a rotational force on the point charges as well. If the flow of electrons is strong enough, it can actually cause the objects to move together while they are revolving, essentially fighting Kepler's laws of orbital motion—which do not include electricity. This concept draws the natural conclusion. The positive charges are attracted to the flow, bind tightly together, and follow the negative charge like a double helix. This phenomenon is referred to as Birkeland currents (1903), named after Kristian Birkeland. It is extremely dumbfounding, but irresistibly exciting that a double helix is a structure so deeply seated in the foundations of Life.

Birkeland Currents can be seen numerous places including solar flares leaving the Sun. The charged currents can also be seen flowing through the center of the Milky Way (pictured below).

The Double Helix Nebulae (Hubble Photo) is explained in terms of electromass on a galactic level. It is no coincidence that this nebula is perpendicular to the galactic plane. The induced magnetic field created by a net flow of charge through the center would account for both the drawing on the previous page of the Nature of galaxies, and this nebula.

Black holes have been determined to be supermassive because of the observations that stars are staying in tighter paths than expected from gravity based orbital motion alone. Einstein's Theory of Relativity locks a speed limit for light, and instead uses time itself as a variable. The tighter than expected orbits along with that accepted math deemed supermassiveness to be a reality. But math doesn't prove concept. Math is extremely important and those tools can be used, but dividing by zero doesn't necessarily mean that the solution is undefined or infinite. Math sets up limits and constraints. It's important to see the truth in numbers; the beauty of regimented guidelines to which one can escape to bring out a definite solution. But also understand that a physical basis must first be made, avoiding assumptions before extrapolations.

As the observations of the center of our galaxy became more precise, astronomers were able to decipher both x-rays and radio waves. It was assumed that the one did not have to do with the other. After all, those two spikes were on opposite sides of the spectrum. Electrons are fast shaking whose action can oscillate at x-ray wavelengths. Their flow and velocity could cause the electrons and their emission to move at speeds, which disturb radio wavelengths as turbulence. The problem is that

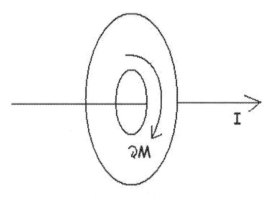

this is not according to currently accepted electromagnetic theory, per Maxwell (1861). Our star is spinning like a football in its trajectory around the galaxy. If other stars are doing the same thing, the magnetic field lines from each of them would cancel (according to the mathematics) in the galactic center, creating no force at all and therefore having no cause for a flow to bind them together. I'm convinced there is a flow, so to 1861 we go.

In the case of two charged iron balls spinning in opposite directions, they still repel by Coulomb's Law, if only a little. They still attract by Newton's Law of Universal Gravitation. But their magnetic fields act slightly different than expected for charges only, because of the mass. The math dictates (and electrical experiments confirm at small scales) magnetic fields cancel between the two moving charges. But since each charge rides a mass it has momentum; charged-masses position themselves slightly wider than expected for charge alone.

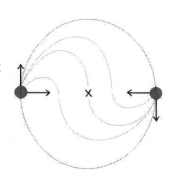

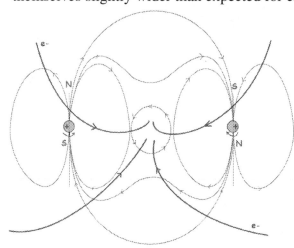

Therefore, the magnetic fields do not cancel. The magnetic fields instead position opposite, but next to each other. This one consideration sets me apart from total submission to Classical Physics as of the late 19th Century, over 120 years ago.[C] Quantum Mechanics can't handle this special case. General relativity alters space and time to explain it. If that one consideration is correct, and the observations in astronomy seem to point to it, our entire paradigm of every field of Physics changes overnight.

With more charged masses positioned around the orbit, the magnetic fields add to make a circle at the center of the orbit. This circular resultant field is the same as a current carrying wire, and so that resultant field creates a flow of current. In a beautiful display of circular reasoning, the net flow of current and this circular magnetic field "binds" the charges in their path. There will always exist

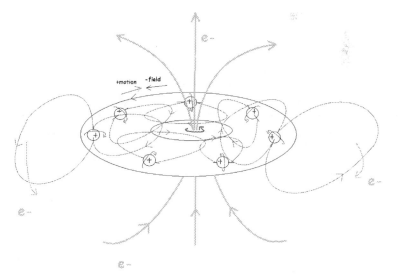

a stabilization distance based on spinning charged masses in orbit, and this flow of charge will be both resultant and attributing. A centrifuge built just right should create or amplify an electrical system, actually generating electricity. It is my educated guess that in the case of groups of charged-masses spinning in opposite directions (from an observer's viewpoint), or in the same direction (from a radial viewpoint), this also creates galaxies without altering reality. It should be noted that proving this takes some of the most difficult, but also most fundamental mathematics in Physics, combining Coulombs Law, subsequently Gauss' Law and Newton's Laws in Orbital Motion. There might be an easier way—read on.

electromass −19

PLASMA:

Atoms were discussed in an early chapter with regard to their emission spectrum. In Donald Scott's, "The Electric Sky," (2006), the reader is introduced to most of the electrical phenomenon in space, and the book inspired me to write this manifesto about the Physics involved. Scott's electrical engineering depiction captures eyes and hearts. To summarize the forms of plasma: Non-emitting, emitting, and by the electric Arc. With high mass objects that have comparatively little charge density, you have non-emitting plasma. When the plasma has little *mass*, but lots of *charge*, you get a glow like a neon light. When there are two high charged plasmas near each other, there could be an electrostatic discharge between them, in the form of an arc. This is seen on arc welders where there is a large current flowing through a small area and heat changes the molecular structure of the metal being welded. Recall the tabletop plasma balls that redirect when you touch them, changing the ground. The beautiful thing about this is that it is scale independent. You may ask, what about the three properties of matter: solid, liquid and gas? The answer: They are all plasma.

[Aurora: Table 16B]

"Seen here over Alaska, auroras are native to the far northern and southern lands. The most powerful magnetic storms can bring auroras all the way to Texas." [Table 20B]

EDM machining caused by a
lightning bolt striking a golf flag.
(source unknown)

There have been observed what are called Fulgurites, deep scars in the ground from
electrostatic discharge machining. At Colfax & 24th street, Minneapolis, MN, 2006,
Lightning scar on a concrete sidewalk is a beautiful representation[/10]

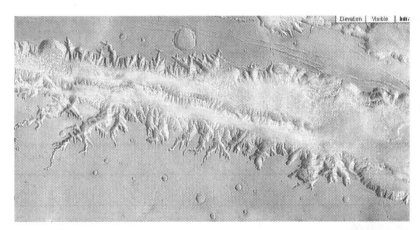

Valles Marineris, Mars

The weld on mars is over 800 times the
size of the Grand Canyon. 277 miles long, ranging
in width from 4 to 18 miles, and as deep as 6000
feet, the Grand Canyon is long since ruled as being
caused by two rivers in ancient Earth, meeting, and
then carving what is now seen today. It's a pretty
convincing story and I don't pretend to be a
foremost on the entire of geology. I would add to
those who disagree about the nature of the Grand
Canyon: Out-dated, unilateral sciences need to
mesh to explain ANY portion of Nature. Let's not
hang on too tight to our guesses from history.
Let's learn from them.

Wait!

Could there be another cause (than the flow of the Colorado River) for the missing material and dendrite-like shape at The Grand Canyon on Earth? [/11]

I'm not the first person to say it—and it's a stretch—but an interesting one to say the least when compared to the scar on Planet Mars—especially now that we have a new mindset. It's a pretty scary and awesome thought that an external source like a passing comet could do such a thing. It makes us puny earthlings pretty vulnerable. Is it worse to know those things are possible, or not to know? Can we develop technology to predict and alter a comet's path and protect our real investment, planet Earth? Or should we cover our eyes and shy away from anything scary? Worse, should we cover our children's eyes and leave it to them unprepared? You're halfway done with this book. I'm going down the rabbit hole; you're welcome to come with me.

When you get a chance to witness lightning, observe the dendritic arms of a typical lightning bolt "searching" for ground, to stabilize the electric potential difference between the sky and ground.

Below: *"This composite image shows the jet from a black hole at the center of a galaxy striking the edge of another galaxy, the first time such an interaction has been found. X-rays from Chandra (left), optical and ultraviolet (UV) data from Hubble (right), and radio emission from the Very Large Array (VLA) and MERLIN (jet) show how the jet from the main galaxy on the lower left is striking its companion galaxy to the upper right. The jet impacts the companion galaxy at its edge and is then disrupted and deflected, much like how a stream of water from a hose will splay out after hitting a wall at an angle." Credit: X-ray: NASA/CXC/CfA/D.Evans et al.; Optical/UV: NASA/STScI; Radio: NSF/VLA/CfA/D.Evans et al., STFC/JBO/MERLIN*

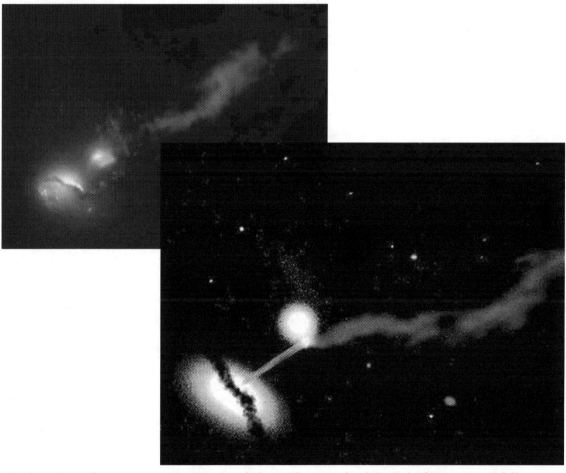

Notice the main stream astronomer and the entire standard model of astronomy fail to properly interpret the cause of the observation, being bound by poor assumptions. To me, it looks like an electrostatic discharge—with electrons flowing at speeds disturbing radio wavelengths. Take a moment to look at the first page of this book—a quote from a simple-minded novice. It would only be fair to try to imagine the mindset of the standard astronomical model pusher: Something profoundly important at any "known" or tangible scale, is *believed* to cancel, *determined* to be bull, and then *stabbed in the face.*

Storms are known to be highly electrical. A cloud builds up a charge on condensed molecules. The condensed molecules drop as rain and the charge buildup is equalized during lightning strikes. What about tropical storms and tornados? The earth scientists are convinced that the temperature difference between the tropical waters and the air causes heat to rise from the water and induces a spin. But salt water is conductive; it holds a charge. What if a tropical storm is just like any other storm, but with the means to transfer the charge more fluidly between the "ground" and the atmosphere?

Left: Tropical Storm, Chantal, 1995 POES AVIRR NOAA.14. Due to the temperature fluctuations of the charged Ocean (temperature excites charged particles), groups of particles (molecules) holding a charge change elevation, and consequently, naturally spin. Yes, heat energizes particles which excite and move. But as soon as their momentum pulls at the bonds between particles—flow of charge—spin.

Individual spin on particles cause electromagnetic fields that tunnel flow of opposite charge between them, as pictured to the right. The Magnetic field goes into the page at the axis of the particle rotation, and coming back out of the page in-between them, causing a tangential (upward) flow of charge. The flow causes more spin. Naturally, the cycle

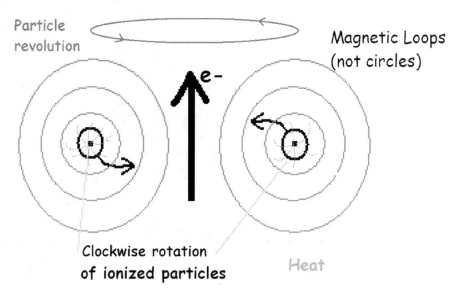

propagates itself resulting in damage and destruction. The particles could spin the other direction and the flow would be downward. The flow may be the positive ions from the sun instead of negative ones from hot water (pictured). The point here is that theoretically, we should be able to adjust the weather patterns ourselves to control climate as well as precipitation. Agriculturally this is crucial; but more importantly, we should be able to steer natural disasters and avoid these tragedies.

electromass −24

Charge is received and bent by the solar electromagnetic field. The Earth's Magnetic Field created by its charged rotation is shaped by the Sun's radiation). Terms like magnetosphere and magnetohydrodynamics become redundant.

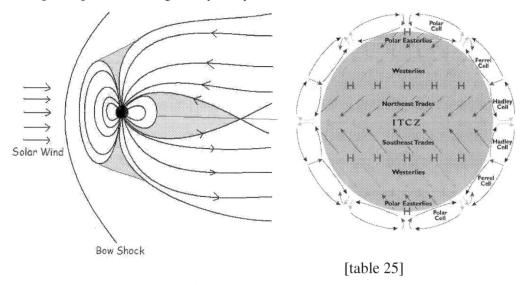

[table 25]

It is accepted that that the trade winds have something to do with the solar winds. It's only natural to suggest that tropical storms are a result of this as well, and not just a heat transfer in the water. The interesting thing about Nature is that all aspects of it must be interrelated. Assigning correlations to aspects of human sociology could be considered either insightful or delusional; but Nature always agrees with itself--even if we can't understand it. Scientists absolutely cannot disregard any aspect of Nature when attempting to describe it. If we can deflect radiation from the sun, we can essentially manufacture ideal weather conditions in more places on Earth.

COSMOLOGY:

The Milky Way Galaxy is said to be of the black hole variety. Observations of stars near its center show perturbations of the stars allegedly caused by an invisible, super-massive hole confusing space and time--some math says so. Our nearest neighbor Andromeda is a Syfert galaxy with a strong concentration of stars at its core.

Parallax angles were the first attempt at ranging long distances without using measuring tape. It uses similar triangles. We witness its effects when we drive down the highway and look off at a distant tree. We seem to be moving slower with respect to the tree as compared to the pavement below. If we pick a point on the roadway, and again at a later, known distance, we can use parallax angles to determine how far away that tree is. The same rule applies for stars, but the accuracy drops very quickly. Over a six month period, you can check the galaxy's position based on its backdrop; assuming the backdrop is fixed (and that's a big assumption). Even our closest neighboring galaxy's distance from us gives us a substantial error when using these angles. But what we *can* observe with accuracy is the observed size in arc seconds when the earth travels over a fixed line of sight. We need to create a vector equation for determining great distance.

electromass –25

The basic idea of the electromass concept came while studying the observations of the most distant galaxies. The goal was to draw a correlation of very distant galaxies—specifically, the size (arc) of distant galaxies vs. their dimness. Astronomers saw a minimum critical point. It was misunderstood as "acceleration of the universally expanding universe." From an empirical standpoint of electromass, there was no reason to cling to an ever-expanding universe, or an accelerating one. The correlation between mass and charge seems reasonable to have such a critical point. It is actually predictable. This thought sent me over the edge; I decided it was too obvious now that the atom, and the atomic forces, can be drawn by virtue of *Electromass*. Let me show you.

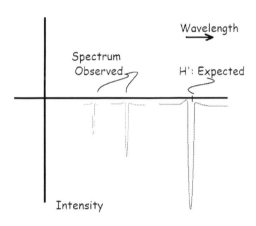

Redshift is a translation of the emission spectrum from where it is normally seen on earth. For example, the Hydrogen emission has a spike at 6600 Angstroms. The sun also has a spike there, which led astronomers to see that Hydrogen is on the surface of the Sun. But for other stars, especially dim ones, the spike for the Hydrogen emission is shifted. As things get dimmer and dimmer, they also get more and more red-shifted. This led to Edwin Hubble's conclusion in the early part of the century, during the Great Depression, that red-shift equals distance, and that was a way for us to calculate distance to a star.

As galaxies got dimmer, the overall diameter of that galaxy also got smaller. The experts proclaimed the entire universe was expanding. Redshift became the vital ingredient for both distance and recessional speed, at the same time, and the result must be the observed dimness. That was the idea anyway. In the 1980's Hubble Telescope was able to capture images so dim, and the size of the galaxy actually started getting bigger. /12

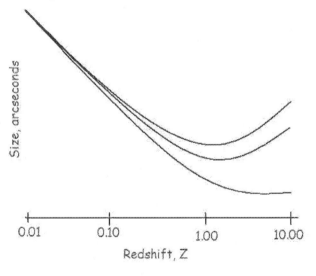

The pushers of the main stream astronomical model didn't flinch. They decided that *dark energy* caused the *accelerated expansion of the Universe*, another physical concept unknown to actual science. They determined that's what made those galaxies so big so fast. In 1998, this was corroborated with a time-dilation of distant supernova. "Since dim supernovae were observed to take longer to settle, things must be going faster now." The astronomers were satisfied that they had figured it all out. Since we can now observe to the extents, let's check all the old guesses (at the door) and take this down logically.

Allow me to offer a different scenario: Size is, I dare to say, a relative term. That is not to be confused with relativistic effects of Einstein and those in his day. The point is that two things can look identical, but have different sizes, and two things can look totally different but share overall size.

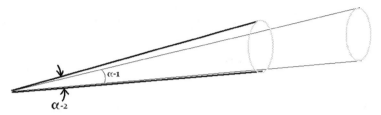

Here is a case where two galaxies could be seen from earth. Observations would indicate that the α-1 arc angle is smaller than the α-2 angle, so the first galaxy would be deemed smaller AND further away than the second—an incorrect assumption in that it is only half right.

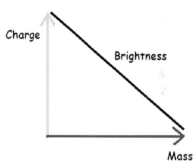

As galaxies get more distant, they get dimmer and their overall size gets smaller—sure—if their masses and charges are similar. However, if there is more charge on a mass, then those galaxies will burn brighter, stars would spin faster, and the galactic radius will be different than the first. If there is less charge on the same mass as the original galaxy, it will be dimmer and revolve in a larger arc. My claim is that there are a few correlations, which are certainly NOT "anti-gravity" or "dark energy." From what we know about mass and charge, my guess is that they look something like the following:

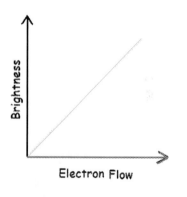

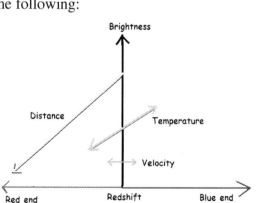

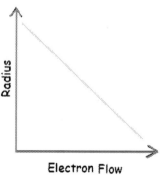

One experiment that seems to affect red-shift is called the Wolf-Effect[5]. It shows that a photon traveling through conductive plasma will lose energy in the form of small quanta, which then heats the plasma medium. We know temperature plays a major role in color and color is seen across the spectrum. This effect, or shift, cannot be ignored.

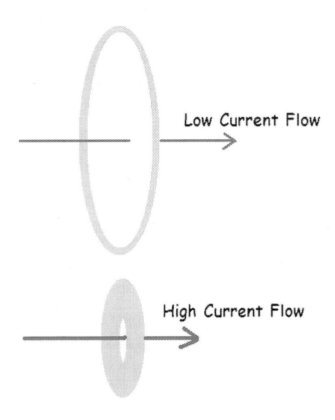

Low Current Flow

High Current Flow

If there are areas in the cosmos that have a greater amount of electrons, then they will flow more towards the galaxies nearby. This will bind the galaxies tighter and force them to burn brighter. If the galaxy has a larger charge density, stars will spin faster, and have stronger magnetic fields and consequently more incoming flow, and more binding. This is counter-intuitive because repulsing "like" charges should get further apart with higher charge; but it is my claim that the resultant, causative, or affective flow of electrons through the center makes it possible. This axial flow is seen in radio and x-rays after long duration exposures (1 week or more), and is studied rigorously along with the phenomenon of quasars by Halton Arp in his 1998 book[/6], "Seeing Red."

Here is an example of how two galaxies can be the same distance from an observer, have the same brightness, but be different sizes. According to Kepler's Law of Orbital Motion, a larger mass system will have higher inertia, and therefore, orbit at a greater radius. If the mass and charge proportions are similar, then the two galaxies would have the same brightness because the individual stars are proportioned equally.

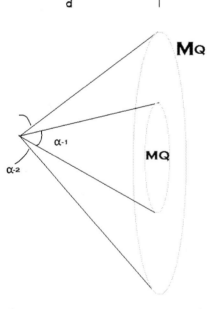

The famous $E=mc^2$ neither accounts for heat build up from a charged conductor ($P=I^2R$), nor this consideration that brightness is effected by the mass-charge ratio or radius of the system. The discovery of pulsars (stars changing brightness) violates the famous equation, and the laws of thermodynamics were used to explain it by "forcing" that the size of the star itself was growing and shrinking—which is irrational.

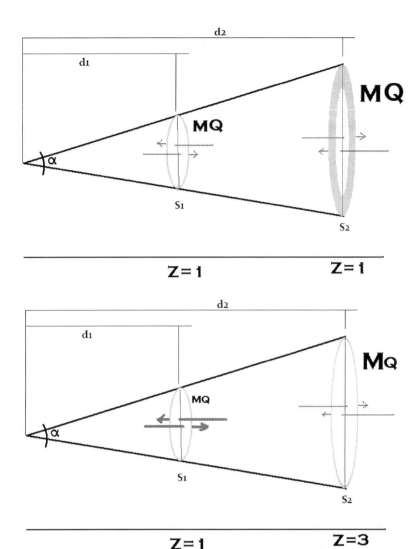

These graphs show galaxies that are observed to be the same size to an observer (α). They can have the same or varying red-shift (Z); the same or varying temperature (in bold); the same or varying distance (d). It comes down to mass to charge proportions on individual stars, and on the amount of free electrons in the area binding them to produce this radius (s) and that temperature. Since Temperature and distance both go like red-shift (and possibly movement), the factor of Z is no longer a useful tool for determining distance without including all of these qualities: Mass, charge, Spin, flow, angle α. Galactic velocity with respect to each other may play a role in Z, but not as an explanation for an expanding or accelerating universe.

Let me say again: all factors for red-shift and brightness (and these seem the reasonable ones) must enter in to any distance equation. Once distance is determined properly, we can track actual brightness vs. galactic radius. We will have a tabulation of mass, charge density, and radius as a function of the electromagnetic spectrum received. It compels one to believe that the equations drawn from that tabulation will match the atom using the same variables, give us a better system to combat and cure disease and work with Nature to provide the best possible situations for human development. Too ambitious? It's time to stop thinking outside the box, and forget the box altogether.

electromass –29

LIFECYCLE OF A STAR:

The following is a collection of conclusions based on visual observations, as well as the fundamental principles of electromass, in an effort to explain the very bizarre. It requires that we skew from the accepted model in many cases, and revise our interpretations; beginning with the collection of charged dust to form nebulae:

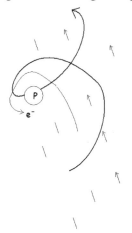
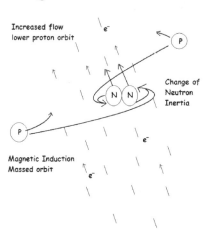

In the above cases, there is attraction of mass and charge, and the drawing in of the necessary components by the same attractions. In the first case, an electron is attracted to a Proton in motion. The proton spins as a result of the electron's approach.

Many elements and molecules are formed by the naturally attracting collisions when trapped in a string of electrons that follow each other as current. When holding a net charge, these molecules will rotate around that string.

NGC 7293: This is also known as the Helix Nebula and is located in the constellation Aquarius, and is one of the closest of all

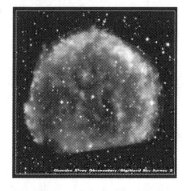

planetary nebulae: Lying at a distance of perhaps 450 light years, it is the only planetary nebula for which a parallax could be obtained by ground-based observations. It is also one of the largest planetary nebulas known: Its apparent size covers an area of 16 arc minutes diameter, more than half of that of the full moon; it halo extends even further to 28 arc minutes or almost the moon's apparent diameter.

Chandra image of Tycho SNR(right). The colors in the Chandra X-Ray image of the hot bubble show different X-ray energies, with red, green, and blue representing low, medium, and high energies, respectively. (The image is cut off at the bottom because the southernmost region of the remnant fell outside the field of view of the Chandra camera) Credit: Chandra X-Ray Observatory/DSS2.

Credit: C. Robert O'Dell and Kerry P. Handron (Rice University), NASA. This image was taken in August, 1994 with Hubble's Wide Field Planetary Camera 2. The red light depicts nitrogen emission ([NII] 6584A); green, hydrogen (H-alpha, 6563A); and blue, oxygen (5007A).

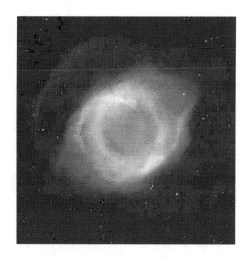

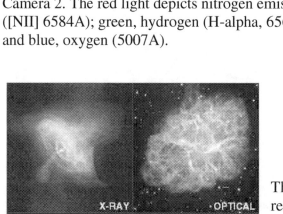

The new Chandra X-Ray Telescope has recorded detailed pictures of the heart of the Crab Nebula (right), first seen on Earth in the year 1054. Here are pictures of the Crab at x-ray (Chandra), optical (Palomar), infrared (Keck), and radio (VLA) wavelengths. [Table 14]

Credit: NASA, ESA, HEIC, and The Hubble Heritage Team (STScI /AURA)

Acknowledgment: R. Corradi (Isaac Newton Group of Telescopes, Spain) and Z.Tsvetanov (NASA)

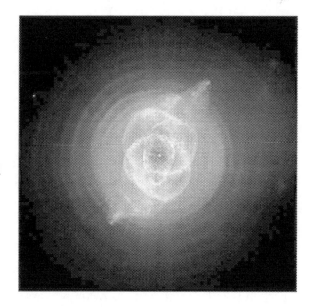

NGC5139

M51 [Table 32]

Once the nebula collect and form the heavier elements present in population II stars, groups of population II stars gather with a natural rotation around each other depending on the original spins of these stars and the surrounding flow. In one case above, stars accrue but with no obvious rotation. This could be due to odd rotational alignments of each star due to a more widely choreographed current flow in the area.

There is a phenomenon that we observe. It is the **Red Giant, White dwarf pairs**. I believe this is what happens when an overcharged star discharges to a nearby dead lump of iron.

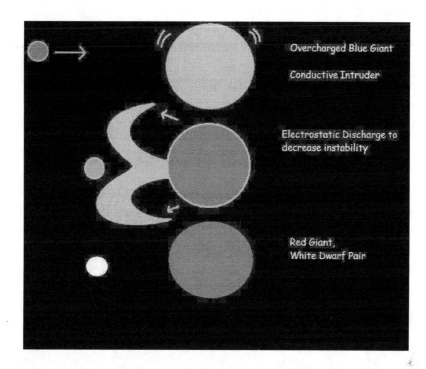

Overcharged Blue Giant

Conductive Intruder

Electrostatic Discharge to decrease instability

Red Giant,
White Dwarf Pair

The accepted model is that a star runs out of hydrogen, grows, turns into a red giant, shrinks, then turns into a white dwarf over millions of years. I do not believe that the white dwarf would explode and create a two star system either. I believe the charge from the giant was transferred to the dwarf by a peeling action described in Scott's Electric Sky[7], rather than explode its component mass. This is a reasonable solution to an observation that red giants often have companion white dwarfs. It is never reasonable to assume that adding "millions of years" will account for observations that we do not understand.

> "The sun will eventually burn up its entire hydrogen core. That will leave a helium ash core, surrounded by hydrogen, and when the hydrogen burns up, the sun will become a Red Giant. Earth will be consumed because the sun will expand to consume the earth's orbit. All of the inner planets will be consumed." - Current accepted model

Already in high school, I began to harbor doubts about this model. Apples to bananas, what do you think? It's ok to doubt the accepted. The standard model is what the combined knowledge accepts. It means we can say, "I think there's a better way."

Supernova events: A combination of electric capacitance and the speed of rotation.

This was discussed earlier. If the whole sphere was charged uniformly, the Electric Field at a point "P," would increase with radius because it would be spinning faster than the center. This leaves the axis of rotation with less of an electric field pressing out. If the sphere was spinning fast enough, the surface charge would become too great for the surface area. At some critical point, there would be an arc discharge at the poles, resulting in a supernova event.

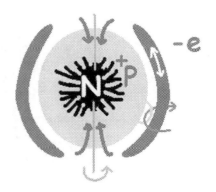

Another thing can cause point P to speed up and result in this supernova event would be a tighter orbit in a galaxy. Tighter orbits are moving faster and have a greater pull. Point P would be moving faster with respect to an outside observer, so revolving faster is just like spinning faster. It is no surprise that observations generally find supernova events are within these tighter orbits in galaxies.

Colour-composite image of the Type Ia supernova SN 2006dr in the spiral galaxy NGC 1288, as observed with ESO's Very Large Telescope. It is based on images acquired through several filters (B, V, R, I and H-alpha) for a total exposure time of 5 minutes. The supernova is the bright object visible below the centre of the galaxy.

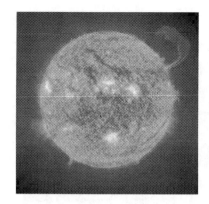

The collapsed core of a supernova contains heavier elements than a Population II star. The other debris from the explosion begins to orbit or ellipse this star, or is shot far enough away never to return. That debris collected to become planets, asteroids, moons, etc. This is what I (and many others) have come to think was the case in our solar system, ~5 billion years ago[B]. It is of the collapsed core of a supernova event that creates the Population I star out of a Population II star—gone awry. The accepted model does not agree with this analysis, in fact it contends that the Population II stars with less density, are older, and that nebulae are remnants. You are the jury.

SOHO/Extreme Ultraviolet Imaging Telescope (EIT) consortium

Our solar system has survived with the sun still burning[B]. Proton's are rolled off, accelerate and collide into a wall of electrons. Suppose: As the core of Population I stars cool, the nuclear reactions become less frequent; they spew out less and less charged Hydrogen and neutrinos; the overall charge of the star decreases; spin decreases; magnetic binding decreases. As with a proton, the star stabilizes at a temperature, and no more protons escape. The nuclear reactions cease. The planets and other debris change orbit throughout this transition based on the constantly changing *electromassive* forces. Its weakened capacitance doesn't create enough of a magnetic field to hold it, and it drifts away from the other positive charges, only to be drawn to the negative charges along the poles. A sheath of current spirals around it as it becomes not a capacitor, but an *inductor*.

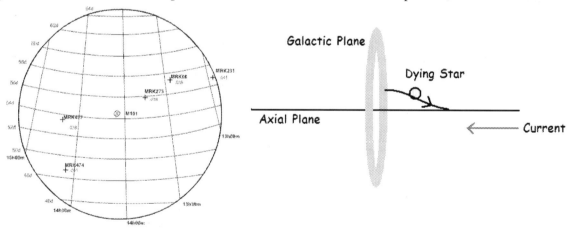

Around the very bright nearby galaxy M101 it is shown that five of the brightest and most intensively studied Markarian (active) galaxies are strung out in a direction roughly NW - SE. Within a *10 deg arc*. All Markarian galaxies brighter than *15.3 magnitude* and *z between .035 and .045* are plotted. [Table 35]

With the strongest Syfert galaxies, these inductors begin to hold too much for the surface area and arc. The flow of heated plasma would shift the observed spectrum around the dead star by the Wolfe Effect. Here we have a star's counterpart: a **Quasar**.

Composite image of spiral galaxy M106 (NGC 4258). Optical data from the Digitized Sky Survey is shown in yellow, radio data from the Very Large Array, X-ray data from Chandra, and infrared data from the Spitzer Space Telescope. The anomalous arms appear as a tilda under x-ray and radio wave emission. The quasars are located axially, centralized, and within the tilda. Credits: NASA/CXC/Univ. of Maryland/A.S. Wilson et al. Optical: Pal.Obs. DSS; IR: NASA/JPL-Caltech; VLA & NRAO/AUI/NSF

Are the highly redshifted quasars part of their parent galaxy? Are they ejected from that galaxy? Do they stabilize the flow received at the galaxy? Do groups of galaxies align along these currents? I believe the answer to all of these questions is, "yes." Halton Arp is the authority on quasar data with syfert galaxies. He has shown galaxies aligning along the same plane. This is impossible for a chance collection of debris from some big bang. It expels the foundation of the expanding universe model. It contradicts the red-shift = distance and recessional velocity. It is our duty to see this openly. Arp, Manuel, Alfvien, Birkeland, Glasser, Ratcliffe, Friberg, Jeurgens, Marmet, Hoyle, Kuroda, Pettengill, Tesla; the men and women who accept electricity in the cosmos, Talbott, Vukcevic, Girart, Steve Smith, Peratt, MGMirkin, Scott, Thornhill and many more. Their work confirmed the concept, and they deserve recognition—for their willingness to approach skeptically—looking to advance the cause of science. Others (who request anonymity) have played a major role discussing the accepted models and to them I am sincerely indebted.

EMISSION:

As stated, light is a spectral signature for an object that is emitting it. The fundamental equation for emission has that same form.

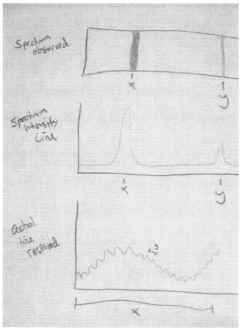

If we take a second look at what we are actually receiving, we see spikes at certain wavelengths. But in reality, the single, comprehensive received line is a combination of each frequency and each frequency's strength. For the equation of electromass to work, all of the signatures would have to be redrawn to this single line. Once this is done, you have deciphered the signature into real terms, and mathematics can integrate the graph as functions of spatial coordinates. Since we have spelled out the existence of a correlation between mass and charge, the equation begins to take form.

The correlation between mass and charge should exist for any spectra. Physicists have spent decades trying to find one by studying Quantum mechanics and atoms. Quantum mechanics has been able to draw atoms, and calculate energies for emission. With electromass in cosmology, the turning point around Z=1 can help draw a better equation for the nature of electromass, and therefore, a better equation for an atom.

For every scale in Nature, there exists a point where the mass dependent attraction due to separation of mass is overruled by charge density, due to separation of charge. Further, the spin of individual mass-charges will define a magnetic field and the distance between mass-charges will be defined by sharing magnetic field strengths and directions. A current flow will also define the bounding radius between those separated mass-charges.

For every scale in Nature, there exists a point where the charge density overpowers molecular structure, and neutron repulsion can be triggered via electrostatic discharge, dissipated in the form of heat energy.

Mass-to-charge ratio and spin on individual stars determines brightness. The flow of electrons axially affects brightness and galactic radius, as well as rotational speeds.

The partial time derivative of the received spectrum is equal to the spatial integral of the sum of the fields (H=ExB·M) with respect to the change in a unit of time.

The line integral of the electromagnetic spectrum received, with a dot product applied to a unit incoming position vector, is equal to a geometric combination of mass and charge enclosed.

electromass –37

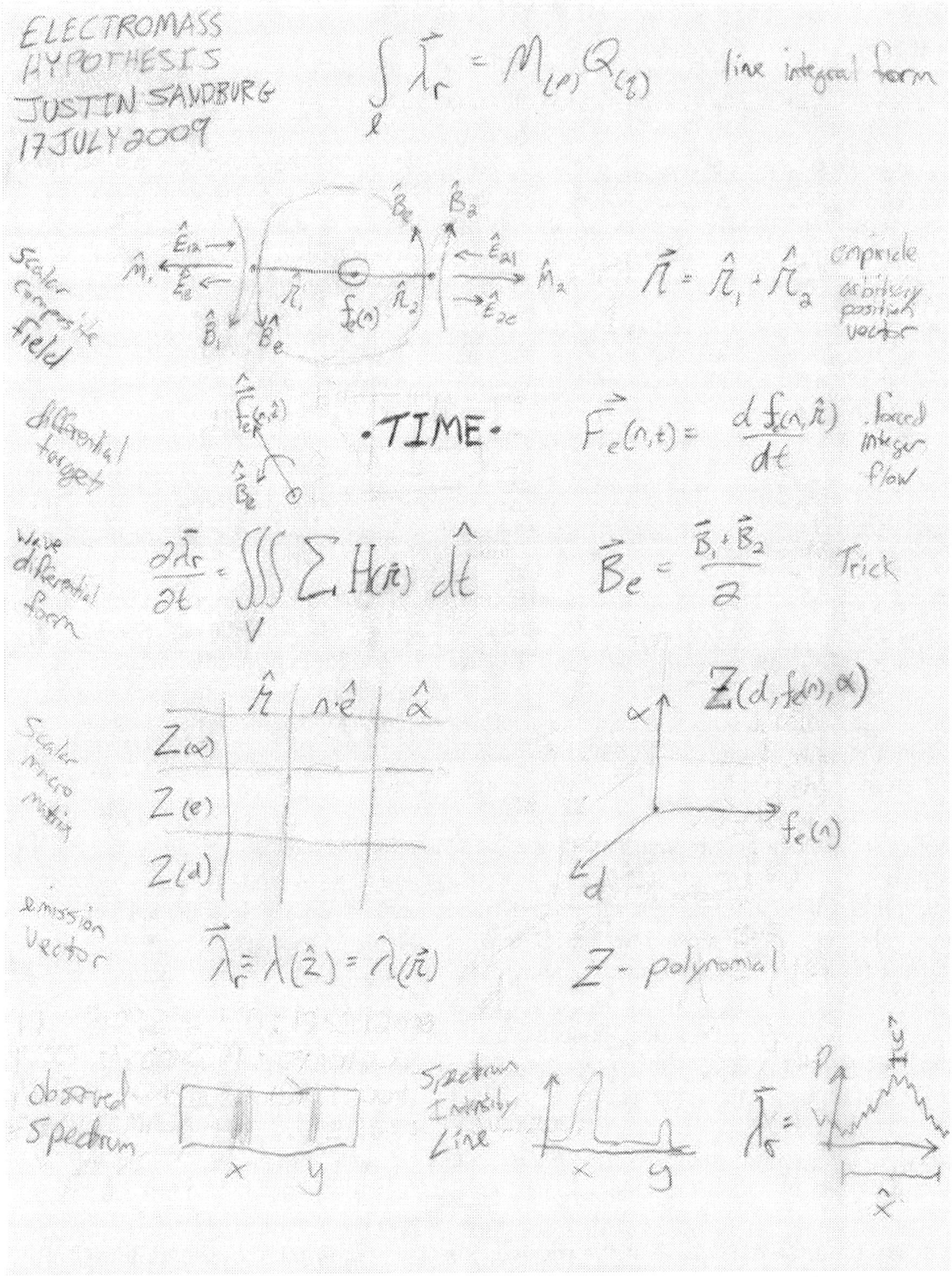

ELECTROMASS
HYPOTHESIS
JUSTIN SANDBURG
17 JULY 2009

$$\int_\ell \vec{\lambda}_r = M_{(\rho)} Q_{(\delta)} \qquad \text{line integral form}$$

scalar
composite
field

$$\vec{\pi} = \hat{\pi}_1 + \hat{\pi}_2 \quad \text{composite arbitrary position vector}$$

differential
integrity

$$\cdot TIME \cdot \qquad \vec{F}_e(n,t) = \frac{d\, f_e(n,\hat{\pi})}{dt} \quad \text{forced integer flow}$$

wave
differential
form

$$\frac{\partial \vec{\lambda}_r}{\partial t} = \iiint_V \sum_i H(\vec{\pi})\, \hat{dt} \qquad \vec{B}_e = \frac{\vec{B}_1 + \vec{B}_2}{2} \quad \text{Trick}$$

scalar
micro
matrix

	$\hat{\pi}$	$\hat{\Lambda}\cdot\hat{e}$	$\hat{\alpha}$
$Z(\alpha)$			
$Z(e)$			
$Z(d)$			

$$Z(d, f_e(n), \alpha)$$

emission
vector

$$\vec{\lambda}_r = \lambda(\hat{z}) = \lambda(\vec{\pi}) \qquad Z\text{- polynomial}$$

observed
spectrum

spectrum
intensity
line

$$\vec{\lambda}_r =$$

electromass −38

ELECTROMASS

$$\oint_\ell \lambda_r \cdot d\hat{\ell} = M_{(P)} Q_{(q)}$$

1

2

$$H_{(\vec{\pi}, n, t)} = \vec{E} \times \vec{B} \cdot M$$

3

$$\vec{\pi} = \vec{\pi}_1 + \vec{\pi}_2$$

4

5

6

$$\cdot TIME \cdot \qquad B_e = \frac{\vec{B}_1 + \vec{B}_2}{2}$$

7

$$\frac{\partial \lambda_r}{\partial t} \equiv \sum_V \iiint H_{(\vec{\pi}, n, t)} \, d\hat{\ell}$$

8

9

10

$$d = \frac{\vec{\pi}}{\sin \alpha}$$

11 $\vec{\lambda}_r(\vec{P}, d)$

$$\lambda_r(t) = \int_0^T \vec{\lambda}_r(\vec{P}, d) \, dt$$

12

13 $\vec{P}_{(P, q, \vec{\pi}, n, t)}$ (energy units per time unit)

electromass −39

TERMS:

The magnitude of the divergence between Physics to date and this particular way of thinking is towering. It's worth looking at the need to revisit the mathematics used and necessary constants introduced over the years, to determine which ones can stay and which ones become obsolete with different mathematics. Ideas like length contraction and time-dilation fall into obscurity by better explaining what are observed to be relativistic effects in space. The introductions of forces of nature in (atomic scale) Quantum Mechanics (Planks constant, quarks) were used as switches, mathematically introduced in order to follow observed changes. Most can be dismissed with better math that doesn't require those switches. The idea of electromass uses variables of light itself to determine properties of the ingredients, so one could actually develop the technology without ever knowing or caring what a proton or an oxygen molecule even looks like-- being only interested in what it does to the spectrum. Even plank's constant angular momentum of an electron, one of the most hardened constants of nature, becomes a burden when we think of only partial changes or proportions of electrons.

~~Aether~~

~~Quantum Electrodynamics~~ (QED) (backwards)

~~Force — weak nuclear~~

~~Force — strong nuclear~~

~~General Relativity~~

~~Special Relativity~~

~~Dark matter~~ (non-emitting charged mass, redundant)

~~Dark energy~~

~~Quantum~~

~~Muons~~

~~Antimuons~~

~~Mesons~~

~~Quarks~~

~~Bosons~~

~~Spinons~~

~~Holons~~

~~Purple Bunnies~~

~~Fermions~~

~~Quantized Angular Momentum~~

~~Speed of Light as a maximum, c~~

~~Supermassive/Superdense~~

~~Anti-matter~~

~~Big Bang Bull~~

~~Acceleration of the Universe~~

~~Expansion of the Universe~~

~~Magnetohydrodynamics~~

~~Magnetic reconnection~~

~~Electric Sun Model~~ (half right)

~~Hydrogen filled sun model~~ (standard model, impossible)

~~Nuclear Fusion Only, Sun Model~~ (half right)

COSMOLOGY RESEARCH
25 JULY 2008

PROOF OF DISCOVERY:
12 SEPTEMBER 2008

CONCEPTUALIZATION:
7 NOVEMBER 2008

PROOF OF CONCEPT:
15 MAY 2009

CONCEPT PUBLIC
29 MAY 2009

ELECTROMASS HYPOTHESIS:
17-24 JULY 2009
25 AUGUST 2009
7 SEPTEMBER 2009

FIRST PRINT
29 JANUARY 2010

CURRENT REVISION:
4 JULY 2010

REFERENCES:

1/ http://education.jlab.org/qa/mathatom_04.html
/2 http://www.answerbag.com/q_view/385527
/3 http://wiki.answers.com/Q/How_many_individual_molecules_make_up_human_DNA
/4 http://www.cdli.ca/CITE/bw_cells.pdf
/5 www.plasma-universe.com/index.php/Plasma_redshift
/6 Halton Arp, Seeing Red, 1998
/7 Donald Scott, The Electric Sky, 2006
/8 Hilton Ratcliffe, The Virture of Heresy – Confessions of a Dissident Astronomer, 2007
/9 Nuclear Systematics, 15 September, 2004, O. Manuel.
"cradle of the nuclides," (The Sun is a Plasma Diffuser that Sorts Atoms by Mass, 2005)
Nuclear Systematics, 15 September, 2004, O. Manuel
Earth's Heat Source – The Sun, Energy & Environment, volume 20, No. 1 2009, Oliver K. Manuel
/10 http://www.celestialmonochord.org/2006/05/index.html
/11 http://azadventuretours.com/COWBOY_S_AND_INDIAN_S.html
/12 http://www.astr.ua.edu/keel/galaxies/obscosmo.html

Tables:

16C. http://www.zam.fme.vutbr.cz/~druck/Eclipse/Ecl2006l/Tse2006l_1640_eit304/0-info.htm
16, 16B. http://www.swpc.noaa.gov/primer/primer.html
20B. http://www-istp.gsfc.nasa.gov/istp/outreach/cmeposter/images/storm1big.gif
25. http://www.newmediastudio.org/DataDiscovery/Hurr_ED_Center/Easterly_Waves/Trade_Winds/Trade_Winds.html, author Tinka Sloss
32. http://apod.nasa.gov/apod/ap060219.html
35. http://www.haltonarp.com/articles/origins_of_quasars_and_galaxy_clusters

Related Topics:

A. Basil Mahon's The Man Who Changed Everything: The Life of James Clerk Maxwell. The first excerpt concerns Maxwell's fluid analogy and comes from Chapter 5: "Blue and Yellow Make Pink, Cambridge 1854 – 1856," pages 59 – 65.

B. There has been a discovery in sunspot cycles, where radioactive decay rates change in Silicon 32, a lightweight dielectric, whose nucleus configuration may change due to a forced flow, similar to a boater being thrown from a turning boat. For heavier and much stronger elements, Polonium 244 or Uranium 238, this decay is impervious to generalized temperature swings and sunspot activity but partially uniform radioactive decay may impact our understanding of the geologic timeline after the supernova event.

C. In August 2009, the word electrogravitics attributed to Thomas Townsend Brown was uploaded to the internet, claiming propulsion speculated in "saucer" technology since the 1950's. This may have previously been classified information and certainly played no part in my research. The concept of electrogravitics fails, because it relies on quantum mathematics at small scales, and suggests that gravity is the cause, rather than it being the inertial momentum (mass) at any scale.

ELECTROMASS AND THE ECONOMY:

It seems inappropriate for a scientist to address the economic question left over from new technology; but it is a profound situation. At what point does technology actually hurt the economy? The system is built for people to pay for services, treatments, and everything in their daily lives. Businesses in turn, provide those services at a cost, and profit. Everyday workers have jobs at businesses to have access to money, to cover these costs. *The stark reality is that most people work just to break even with the costs incurred for the simplest efficiencies—lights, water, heat, shelter, transportation, security, and food.* For the average person, these are the necessities, which encompass their entire lifestyle. There is not room for luxuries. They struggle diligently at their profession, doing what they're told, and hoping that they make enough of a difference to get noticed by their superiors. Because that's half the battle, right? Jobs? The other half is making money—isn't it? Does this "human energy" maximize efficiency, or profits, or neither?

Education is a focus on any civilization. But questions lead to learning, and learning leads to technological growth. Advancements make older processes obsolete, and greater efficiency reduces jobs. With certainty, technology will ruin an economy based on jobs and making money. It is inevitable. The question is do you stop technology or change the economy?

Bio-feedback machines exist today; but lack computational power to be supremely accurate. If a $40 dollar electrical test could identify illnesses, then what do we do about all the doctors who run expensive tests to diagnose them; and what about the prescribed medications that are formulated from complex processes from very profitable companies? Would a doctor go so far as to diagnose someone with an illness for the sole purpose of selling more pills? Would a company lobby to the doctor and offer trips, vacations, or bonuses for higher sales? If it was that simple--if technology could actually outperform educated humans, then the job loss would never be recovered.

Most "family" doctors (those who do regular check-ups) will be transformed into nutritionists. They will have to really know their bio-chemistry from the ground up, and know which supplements and which foods help to build, say, liver function; or kidney function. Here, education plays a key role in which doctors survive and which doctors quack. Pharmaceutical companies will have to develop perfect strains with less and less side-effects, because a nutritionist can use a method of vitamin and herbal supplements to build function, rather than risk those functions. Nutrition costs a lot of money. A person could spend $250 per month for six months or more on vitamins to get their body into alignment. The poor have little choice in the matter when it comes to foods that are healthy verses foods that will fill up their family--nor can they justify a $60 bottle of fish oil compared to a week's worth of food. If health costs were made efficient, then people would not need to pay thousands of dollars for insurance in order to cover the cost of expensive treatments.

Surgeons, plastic or cardiac, and all specialized doctors survive. But a heart surgeon must realize that in a Utopian society, everyone will be healthy enough and know enough about their bodies to avoid heart trouble in the first place. Those surgeons should, more than most, hope that their job becomes obsolete. A knee surgeon would not walk around with a sledge hammer in order to get more clients. What about people who really like and can't help but crave grease and salt? Well, certain doctors will enjoy them; certain companies will not go out of business because people have the freedom to eat what they want. A poor diet will lead to heart trouble, so why should healthy people pay insurance for trouble they will never have? These bio-feedback tests are so accurate (and non-biased) that a person could pay insurance based on their personal "risk." People who continually show dehydration and laundry lists of other poor function will pay more for insurance, because they will be the ones who will need insurance to help them when their body fails. An alcoholic will show a stressed liver in their bio-feedback, and should pay a portion of insurance premiums to treat the eventual effects. On the other hand, an individual who eats a balanced diet, exercises, and does everything right to keep their body functioning like a well-oiled machine--the electrical test would show that. Those people should pay less for insurance, giving them more money to put into the healthier foods. As people get better at taking care of their bodies, then they will get better scores and their premiums will go down.

What if there was a means of producing energy that was extremely efficient? Page 19 of this book alludes to safe electricity generation machines on-site. We may altogether reduce the need for monthly electric bills. We may not need to stick to oil drilling. Some alternative energy sources are better than others. The economy in general will increase when people shift their income away from pointless energy bills that strangle their livelihood. They would have more to do with what they want, freeing them of a corporate burden of dependence.

Transportation is as fundamental as communication. The world is large and transportation gives us the ability to see many places in a single lifetime. Costs of transportation would be limited to the research and development of using new technologies to have cars that drive with no charging, no pumping, and only run out when the parts run out. But even transportation companies have to make money. A car manufacturer has to weigh material properties and life vs. cost, and people buying those cars realize that. If there was an engine that needed no maintenance, would repair shops discourage the subsequent job loss? Jobs are made and broken based on providing a service. It's possible that one day technology could become so advanced that repair was unnecessary. The argument is whether or not this is a bad thing. The assembly line took out many unskilled jobs, and provided cars faster and cheaper. This was great for business and after 100 years, nobody argues the job loss. Imagine that on a higher level; educated jobs, doctors, being replaced by a machine. Yes, this is good for business but as technology creates durable machines that do everything, nobody will have jobs. Corporations would be the only ones who would succeed, and the population they depend on as consumers would be turned to poverty. The extreme is zero taxes, anarchy, and corporate free-for-all. There has to be a government response from the highest level.

The question of the century is: How do we make money for the economy to function? The trick would be to use the good parts of each type of governance and business sense, to maximize the economy for all, and provide incentives and rewards for education and health, rather than profits and the big squeeze. *The stark reality is that most people work just to break even with the costs incurred for the simplest efficiencies— lights, water, heat, shelter, transportation, security, and food.* The goal with new technology is to reduce those bills on the consumer rather than widen a profit margin. The solution to job loss created by efficient technology requires a fundamental change in mindset.

Welcome to the Electromass Revolution.